IMAGINATION
MY DREAMS PAINTED WITH A.I.

STRAVARD® LUX
PUBLISHING & DISTRIBUTION CO.
www.stravardlux.com

Copyright © 2022 by Yvette Kendall

All rights reserved. IMAGINATION, My Dreams Painted with A.I. belongs to the author, Yvette Kendall. The copyright of its texts is owned by Yvette Kendall. No part of the texts in this publication may be reproduced, distributed, or transmitted in any form or by any means, including photocopying, recording, or other electronic or mechanical methods, without the prior written permission of the publisher, except in the case of brief quotations embodied in critical reviews and certain other noncommercial uses permitted by copyright law.

For permission requests, write to the publisher,
addressed "Attention: Permissions Coordinator," at the address below.

Stravard Lux Publishing & Distribution Co.
3780 Old Norcross Rd.
#103-114
Duluth, Georgia 30096
www.stravardlux.com

ISBN: 978-1-7371440-2-1 (Paperback)

TEXTS HAS BEEN WRITTEN BY YVETTE KENDALL
IMAGES CREATED BY ARTIFICIAL INTELLIGENCE

PREFACE

I am excited to present this book to the world. With that said, I want to assure you that there is nothing life-altering inside. I made this tabletop book out of pure whimsy, beauty, and curiosity. It's a combination of the past and the future coming together gloriously. The images inside have been created by AI. However, the words that anchor them to the pages are mine.

The visuals speak for themselves. I have often stated that "AI has found a way to become a representation of our dreams and nightmares." I have struggled to communicate what I see when I'm asleep. Despite my efforts, I didn't have much success until now.

AI can extrapolate these details with an accuracy never seen before. Over my career of writing Science Fiction and dabbling in Comic Books, I have yet to put ink to paper and manifest such bizarre and fantastic visual concepts.

As you turn the pages, you will find a little bit of everything from the weird to the stunning. As mentioned earlier, it's just a book of compelling things created with an unknown and silent partner.

Yvette Kendall

Imagination, My dreams painted with AI. , Yvette Kendall

~ This is one of my favorite images. My keywords were simple enough, but the program gave me more than I could have ever hoped for.
She's hauntingly beautiful.

The day had arrived without fanfare, crowds, or birds of song. Abbalith cried as she slowly put on her wedding dress and walked through the town and up the mountainside. She stood in wait for a love that would never come. She used a sharpened branch to slit her wrists in one final act of desperation. Looking down from her heavenly perch, Hera took pity on Abbalith and froze her into a golden statue. Her beautifully-bloodied dress she made into a train of blossoming red roses.

~ Yvette Kendall

Imagination, My dreams painted with AI. , Yvette Kendall

~ As you view these stunning images, you will learn that I have a great love of butterflies. I also love to combine them with the colors red and gold. Here is an excellent representation of them seamlessly merged together by AI.

Every one-thousand years a gift arrives that foretells the destiny of the next. We never know the day, time, or place, but when it arrives, it is clear that a death of an age has come and the birth of another is at hand. Our gift is here.

~ Yvette Kendall

Imagination, My dreams painted with AI., Yvette Kendall

~ Who knew that Sir Darth Vader had a softer side? A Sith Lord touched by love and a little Da Vinci. Keep looking, there's more of Darth to come.

Even in the darkest heart can grow a bouquet of love that can be seen from the farthest corners of the Universe.

~ Yvette Kendall

Imagination, My dreams painted with AI. , Yvette Kendall

~ Ahhh.....This beauty is titled, *"God Watching the Last Supper"*. Ever since I wrote my first SciFi novel, The GOD Maps; I have been obsessed with divinity materializing. My keywords again netted me something spectacular!

Judas looked up and saw God in a vision but believed it was wine that demented his perception. He thought, "Surely if God is in this room, then he would cease me to be right where I stand." God looked down at Judas and said, "Nothing goes unseen to a light that snuffs out the darkness. As wretched as you are, your part must take to the stage. Nevertheless, you need not fear because your time is soon at hand."

~ Yvette Kendall

Imagination, My dreams painted with AI., Yvette Kendall

~ This one is so surreal.
He is a Minotaur with gilded armor dying in a maze.
To say he is beautiful is an understatement.

The Sun rose on the darkened maze as Palaxese took his last breath. For centuries he reigned over the stepstones and tunnels to the Afterworld, but no more. Overnight Palaxese had a visitor. It was a girl child who had hair of blonde and eyes of blue. Alas, this was no child, no child at all. In deception, she came wielding the power of Zeus through the tips of her little fingers. The girl smiled, and looked him in the eyes. Then she struck him down. She said, "I came to clear the way. The Afterworld no longer needs a guardian. I will lead the hordes of the damned myself."

~ Yvette Kendall

Imagination, My dreams painted with AI. , Yvette Kendall

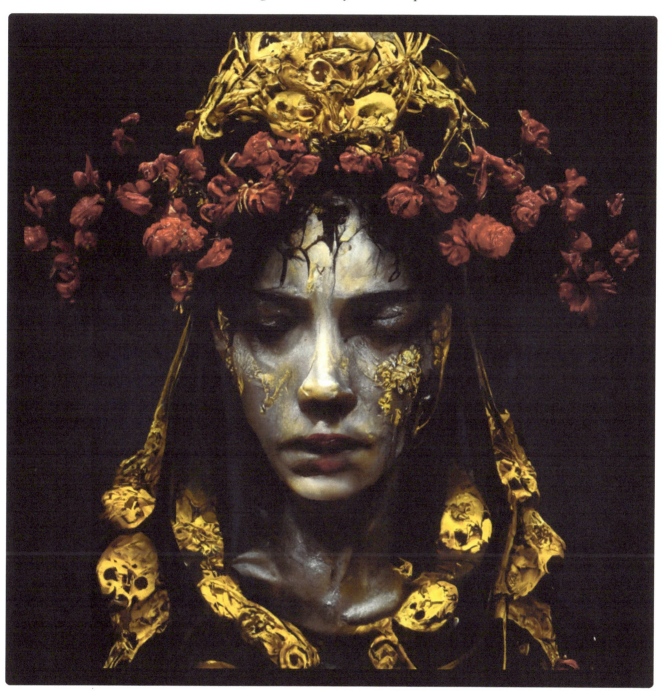

~ How can anyone that cries golden tears be sad?

Pain was all she had ever known. Mayla was installed in the middle of town like a beautiful fountain. She was hoisted onto a pedestal to live day in and day out without care. People came from miles around to make her cry. Children taunted her, men molested her, and women struck her. The cruelty made no matter. Mayla understood that her life was not her own. So there she stands, crying tears of gold. In her moments alone, the tears continued to fall. Because deep in her heart knew that she would never, ever leave this place.

~ Yvette Kendall

Imagination, My dreams painted with AI. , Yvette Kendall

~ I find that there's beauty in the macabre. My keywords were Lucifer and Butterfly, the AI did the rest. STUNNING!

There is beauty in everything, no matter how ugly. You need only to look past its gruesome phantasm to find the breathtaking wonders within.

~ Yvette Kendall

Imagination, My dreams painted with AI. , Yvette Kendall

~ If you ever imagined what the last supper would look like if Jesus was Japanese. Well, wonder no more. Love it!

Junsuke took stock of the room as he analyzed the men in attendance. His heart was filled with love and hope, but his mind warned him of impending doom. He said, "By noon in the morrow, one of your blades will be stained with my blood. Let it be known that a Samurai with no honor is a Samurai with purpose. Now let us drink to this day, for it will happen no more."

~ Yvette Kendall

Imagination, My dreams painted with AI. , Yvette Kendall

~Young, Gifted, and Black....And that's a fact.

Will there be Black people in the future? Why yes, it will! We have been here since the beginning of time, and we will be here when time ends for us all.

~ Yvette Kendall

Imagination, My dreams painted with AI. , Yvette Kendall

~It's all about the butterflies.
I have created a series of throne chairs reimagined. Thinking outside the box takes your mind to new heights. It's bizarre and elegant at the same time.

The little boy asked the Queen, "Why do you sit on a throne made of butterflies?" The Queen responded, "It's to remind me that change only matters to others if they can see the beauty in it."

~ Yvette Kendall

Imagination, My dreams painted with AI. , Yvette Kendall

~Introducing the King and Queen of an underground kingdom.
May they reign forever!

After their marriage, a jilted witch cursed the pair to live underground for the rest of their days. Although saddened by their sentence, the King and Queen made an underworld dominion that would rival the most luxurious of nations.

~ Yvette Kendall

Imagination, My dreams painted with AI. , Yvette Kendall

~ "Game of Horns" anyone?
This throne is comprised of antlers from any and every wild beast that roams. This photo barely does it justice.

It won't be long now. Plans are being made, and traps are being set. Whilst you slumber in your beds made from the feathers of geese and the hides of every wild thing. Just know that the day is almost at hand where the beast will have their day!
~ Yvette Kendall

Imagination, My dreams painted with AI. , Yvette Kendall

~ Astronauts on a planet of flowers.

This was the last planet on the list. We have, maybe foolishly, decided to abandon Earth altogether because of it. Although we are not prepared to stay, we are doing so anyway. Only Heaven knows what we may encounter down the line. However, the sheer beauty of this planet makes all potential calamities worth the risks.

~ Yvette Kendall

Imagination, My dreams painted with AI., Yvette Kendall

~ Darth is back!
Now I've imagined him as a priest.
I have many thoughts on this.

In a land far, far away there is a priest with the ability to do miracles. Many of his parishioners say that he has gifts from the Gods, but others say...nothing at all. His eminence teaches with an iron fist and an equally devastating searing wand. For it is understood that he detests anyone who has a lack of faith. He makes it known that it will not be tolerated.

~ Yvette Kendall

Imagination, My dreams painted with AI. , Yvette Kendall

~ Pondering reality and the creation of it all.

Not many get to see behind the velvet curtain of life. But for those that are blessed to do so...the view is quite amazing!

~ Yvette Kendall

Imagination, My dreams painted with AI. , Yvette Kendall

~ Beware of the pretty woman with the forked tongue. I enjoyed crafting her with my words. She's saying so much without saying anything at all.

Anassa was extremely pretty, but she would never, ever speak. Whenever someone tried to engage her or if she was asked a question, she would only nod politely with a smile on her face. Some kids at school took her prettiness and silence for arrogance. One day, a group of them decided to surround Anassa and aggressively taunted her until tears. They chanted, "Speak...Speak...Speak!" Anassa grew so angry that she opened her mouth to scream, and a horde of serpents roared out her mouth and struck the children dead. As she looked down at the mess littered at her feet, she sucked each Serpent in one by one. Anassa politely closed her mouth and walked away with a smile on her pretty face.

~ Yvette Kendall

Imagination, My dreams painted with AI. , Yvette Kendall

~ Oh...Did I mention, there are ZOMBIES!!!

Although Delilah wasn't the same, she still wanted to be pretty. The rot had set in months ago, so she decided to free her hair into the biggest afro anyone had ever seen. Delilah hoped that its massive hood would help to cover her decaying face. She also found a few roses to finish the job, but eventually, they turned as black as her hair. As Delilah wandered the streets looking for people to eat, she saw an old carnival that boasted one last yellow balloon. She thought, "That's it! This will finish my look off nicely. I'll be so cute that the people won't mind that I'm having them for dinner."

~ Yvette Kendall

Imagination, My dreams painted with AI. , Yvette Kendall

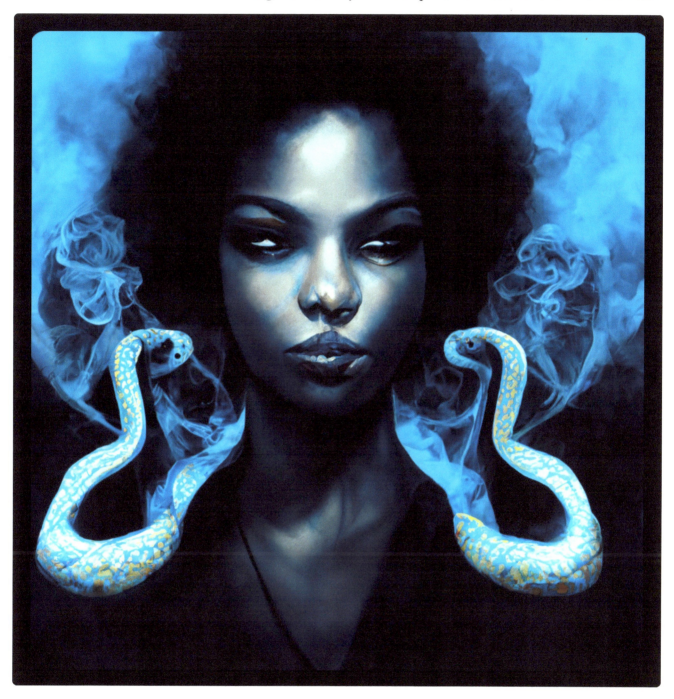

~ The keywords were: Beautiful Black Woman, Blue, Smoke, Medusa.
The AI honored me with a fairytale of an image.
Her name is "Mayko Lox"

Mythology exists everywhere, and Mayko Lox is proof of that! A beautiful Black Woman that commands reptiles with her ancestral powers should not be odd. As the story goes, the Serpent was an ultimate source of wisdom. What you do with those whispered secrets is up to you.

~ Yvette Kendall

Imagination, My dreams painted with AI. , Yvette Kendall

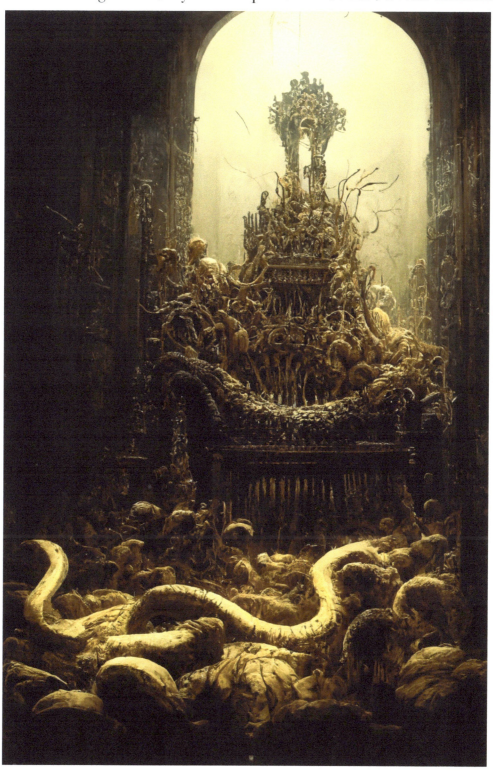

~ This Kingdom rules everything!

The funny thing about power, if you wield it improperly...everything you rule will die.

~ Yvette Kendall

Imagination, My dreams painted with AI. , Yvette Kendall

~"It's a nice day for ah Black Wedding."

Mrs. Moore had a wicked sense of humor. She often cited, "We are born to die, so why not dress for the occasion?"

~ Yvette Kendall

Imagination, My dreams painted with AI., Yvette Kendall

~ The Golden Skeleton"

Last night a nightmare came to get me. Before I could run away, it said, "Your sins flow down your leg to hide in the dirt. But it never leaves, not really. Like a cow chewing its cud, we go back to it. We go back to the wicked fruit to feed our tired souls. Like a dog licking up his vomit, we will give darkness a chance to live another day."

~ Yvette Kendall

Imagination, My dreams painted with AI. , Yvette Kendall

~ This is a world that I want to live in.

Have you ever seen such a sight? Wait, there it is again. The birth of seconds, minutes, and hours. Yes, everything is born, even time.

~ Yvette Kendall

Imagination, My dreams painted with AI. , Yvette Kendall

~ Nebulous Oil Slick

Being black is the biggest conundrum that has ever existed. We seem to hover without roots in spaces that won't allow us to plant seeds.

~ Yvette Kendall

Imagination, My dreams painted with AI. , Yvette Kendall

~ A Xenomorph with a little flare.

"I don't know what waits for us beyond the stars, but whatever it is, it has to be more forgiving than Humans. I'm willing to take that chance."

~ Yvette Kendall

Imagination, My dreams painted with AI. , Yvette Kendall

~ Fly Zombie King

He ruled in life, and now he rules in death. After adjusting his crown, the Zombie King had just one question for his on-looking subjects. He turned to the side, cracked a smile and asked, "Do I look good"?

~ Yvette Kendall

Imagination, My dreams painted with AI. , Yvette Kendall

~ The "Predator" in year 3032

In the future, even monsters will be enchanting.

~ Yvette Kendall

Imagination, My dreams painted with AI. , Yvette Kendall

~ The Steam Punk "Predator"

What can I say? It starts off fun in the beginning. A sweet art program that everybody loves. It gets billions of users creating content everyday. Then one day, this guy shows up at you door with bad news...really bad fucking news.

~ Yvette Kendall

About the Author

Yvette Kendall is an American Author from the south side of Chicago, Illinois. Kendall is the author of The GOD Maps, Volume One, the first installment in the Trilogy. Because of her specific writing style, Kendall went a step further and created a new SciFi sub-genre called Biblical Futurism. The genre has attracted a whole new audience to its unique category. Kendall recently signed an Option deal to turn The GOD Maps into a TV Series of the same name.

Other Titles By Yvette Kendall:

The GOD Maps, Volume one
Horrorgraphs and Other Short Creepy Stories
A Zombie For Mommy!

Upcoming Titles:

The God Maps, Volume II "Rogue Heaven"
The Revelation Activation
A Werewolf For Mommy!

Titles are available online at all major distributors like Amazon, Barnes & Noble, and many more.

Published by Stravard Lux Publishing and Distribution Co.

www.StravardLux.com

CPSIA information can be obtained
at www.ICGtesting.com
Printed in the USA
BVHW020056011022
648389BV00002B/40